Puff
Saves Paradise

"In loving dedication to Puff, Brandy, Sunny the Duck,
Jax the Cat, Crispy the Butterfly, Polo and King."

"A very special thank you to my 2nd grade students
of 2022 who also helped inspire this book."

Brittni Friedlander
Email: Puffsavesparadise@gmail.com
Instagram: @full.steam.ahead

FULL S.T.E.A.M. AHEAD
BOOKS BY BRITT

Written By :
Brittni Caitlin Friedlander

Illustrated By :
Naidielee DL. Laquindanum

Puff
Saves Paradise

Hi there! Aloooooha! I'm a pup named Puff.
I'm a Tibetan Spaniel.

I live on the beautiful island of O'ahu Hawaii,
where the palm trees sway and the turtles lay.

During the day, I go to STEAM Academy. S.T.E.A.M. is an acronym. It stands for science, technology, engineering, arts, and math. Here, we find solutions to solve real-world problems happening today.

I love my group of friends at STEAM Academy. There's Brandy the Bichon, Jax the Cat, Crispy the Butterfly, Pitbull brothers Polo and King, and Sunny the Duck. Together, we make a great STEAM Team!

After school, my Mom and I like to take long walks on the beach's shoreline and play in the sand!

Some days, I like to take my drone
out and fly it over the ocean,
as far as the eye can see.

One day, when Mom and I went to the North Shore after school, I took my drone out and flew it high towards the ocean's horizon.

Out there I saw something
that surprised me...

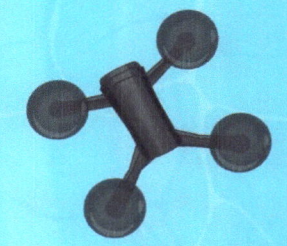

It looked like a baby green sea turtle caught in a fishing net and trash debris.

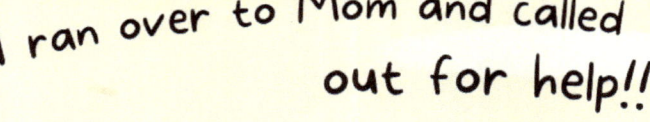

I ran over to Mom and called
out for help!!

Together, we paddled out to find the
sea turtle tangled in a fishing net and
trash <u>debris.</u>

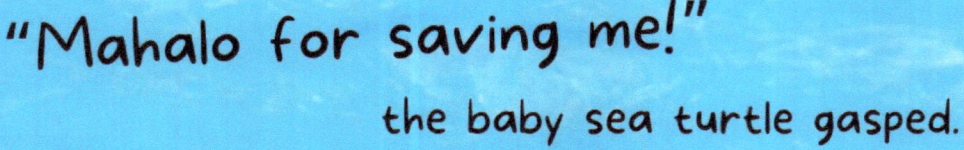

"Mahalo for saving me!"
the baby sea turtle gasped.

Once we brought him to shore,
I asked the green sea turtle
what his name was. He replied,

"My name is Kuhina,
I'm an ambassador of
the sea."

Kuhina was badly tangled and looked quite sick,
possibly from being caught in the debris.

Mom and I knew exactly where to take him for help. Together, we brought Kuhina to Mālama i nā Honu. An organization known for protecting and <u>assisting</u> green sea turtles who need help.

There, they assured me that I could visit Kuhina everyday until he was brought back to health.

The next day at school, I decided to tell my friends at STEAM Academy what had happened. To my surprise, they didn't realize the seriousness of the pollution problem.

That's when I realized: We need to find a solution to bring awareness to this pollution!

I went to my favorite teacher, Miss Atud, and asked her what we could do. Miss Atud was a S.T.E.A.M. educator, which means she combines science, technology, engineering, arts, and math to find solutions.

I knew she would know what to do.

Collaboratively, we came up with a plan to raise awareness. Combining all aspects of S.T.E.A.M., I sketched a plan, and my friends came to help.

Over the next few weeks, we collected over a thousand pieces of trash pollution on the beaches, sidewalks, and streets!

Using the plastic bottles, straws, glass, and fishing nets we'd found, we put our plan into action. We needed to save our paradise.

Together, we decided to create a sculpture of Kuhina that would represent the mass amount of litter we had collected! We got to work right away.

During this time, I visited Kuhina every day after school at his recovery cove. I asked his opinions about our plans and goals.

mālama i nā honu

Because after all, teamwork was the most important part of the process.

Kuhina was so excited to hear about our plans, and he was honored to hear he was at the center of it all.

Gradually, Kuhina got better and he was able to be released back into the ocean. Before leaving, he asked to see the unveiling of our sculpture.

Parents, educators, and community members alike came to see the wonderful creation that members of the STEAM Team made with our own two hands (or for some of us-paws). It was an eye-opening event for everyone!

Everyone was able to meet Kuhina, too.

From there we said our goodbyes to Kuhina,
but we promised we would always be friends.
"A hui hou," he said. Until we meet again."

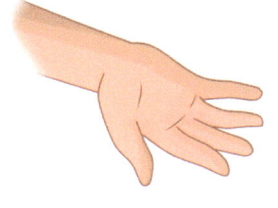

Soon after the unveiling, we saw progress in plastic usage and littering across our paradise. Sunny's dad was a grocery store owner. Their store switched from plastic to paper.

Jax's Mom owned a boutique. She started selling only refurbished furniture and recycled clothing.

As for Mom and me, we became regular volunteers at Mālama i nā Honu! We educated others on green sea turtles and how pollution affects us all.

Even today, our sculpture still stands at the academy as a reminder to keep our environment clean.

And what happened to Kuhina you may ask?

It's been many years since all of this happened, but every so often we catch up where it all started—at the edge of the ocean's horizon.

And that's how we saved our paradise.

Glossary

Acronym: an abbreviation formed from the initial letters of other words and pronounced as a word

Solutions: finding answers; solving a problem or dealing with a difficult situation

Debris: scattered pieces of trash

Ambassador: an official representative

Assisting: the act of helping

Awareness: knowledge of a situation

Collaboratively: two or more people working together

Opinions: a view or judgment formed about something

Refurbished: to redecorate or to fix something

Environment: the surroundings of where a certain person, animal, or plant lives

Reading Comprehension Questions

1. Can you see each aspect of S.T.E.A.M. represented in this book?

2. If so, where can we find each one?

3. Are there other ways you could bring awareness to a pollution problem at your school?

4. What other ways could the students of the STEAM Team have solved this problem in unique ways?

5. How many times is Crispy the Butterfly seen?

6. How many turtles are there on page 6?

7. How many turtles are there on page 19?

8. How many turtles are there altogether on page 6 and 19?

Check out the real life characters behind
Puff Saves Paradise

About the Author

Brittni Caitlin Friedlander grew up on the beautiful island of O'ahu, and is currently a teacher and a first-time author. She has a passion for S.T.E.A.M. curriculum and empowering the next generation through education. As an animal lover and foster volunteer, she's had the privilege of rehabilitating animals of various backgrounds--some of which have inspired this book. With this story, she hopes to educate children on the problems of pollution using S.T.E.A.M. principles in order to create innovative thinkers, creative learners, and community contributors through PUFF SAVES PARADISE.